Dopamine: The Action of Life

3rd Edition

Rowena Kong

2024

First Printing: 2021

ISBN: 978-1-998152-85-8

Contents

Preface

If you have been regularly following popular neuroscientific articles in the media, then dopamine is definitely no stranger to your memory. From Parkinson's disease and schizophrenia to being touted as the "pleasure" neurotransmitter, the general public in this 21st century has never been more acquaint with this multi-faceted working component in the brain than before. Yet, the overpopularity of such association may only represent just 1% tip of an iceberg. For those who are still not too familiar, dopamine is not only active in the normal functioning of our central nervous system, which consists of both our 3-pound heavy brain and the spinal cord which extends down our back, but also in the peripheral nervous system which is widely distributed throughout our body. We are, definitely not just our brain, as most people would like to assume, though it consumes as much as 20% of our body's glucose resource(so much for the sensationalism promoted by popular media). As easy as we and brain are in succumbing to heuristics, we tend to make light of and forget that the other 80% of glucose is still a major proportion of the whole for the body's peripheral needs, which are required to work in concert with the brain's orders.

The truth and fact is that dopamine and its 5 subtypes of receptors, labeled as D1, D2, D3, D4, and D5, are constantly being produced and expressed on various cells of our major body systems, e.g. cardiovascular, digestive (enteric nervous system), immune and renal, to name a few. It is not just our brain or cerebral system which requires this "multi-functional" classical neurotransmitter to get us thinking and functioning in our everyday lives. Our nervous system is actually more encompassing than previously presumed that we literally cannot live without it. At the same time, there is a reason why the popularity of dopamine is enduring. I doubt if its ability to induce a sense of pleasure from our brain is the only one. We need it to combat Parkinson's disease and to ensure that our kidneys are regulating blood pressure efficiently (refer to later chapter on the renal system). Like other classical neurotransmitters that are needed in considerable amounts by our bodily functions, a disturbance of its production, release, activation and inhibition is bound to push our health to the tipping point. Having said, I hope you could begin to appreciate the necessity and benefits that our neurotransmitters and the various categories of our nervous system bring to the stability of our health and the pleasures of life, be they dopamine or not. Such is the aim and priority goal of me in writing this book. In many ways, neurotransmitters colour our perception of the world, without such everything would be a black and white vacuum of inexistent meaning. Amongst the additions in this new edition include an expanded focus on a described novel and rare symptom of "visual broadcast" in schizophrenia, of which chapter 3 attempted to discuss preliminary explanations for such an atypical manifestation in a patient. In doing so, such discussion brought up associations with the concept of stress-induced psychosis and inevitably the etiological significance of both dopamine and cortisol for this condition.

Chapter 1: Reward and Positivity

Most of us would likely remember that rewarding feeling we experience each time we receive praise from others or get to own an item on our top favourite list. While it is not baffling for us to associate positive aspects of speech and possessions with ourselves, the neural correlates of this relationship are only beginning to be unraveled. In a recent research finding published in Psychological Science, Dutcher et al. (2016) reported that making judgments while self-affirming led to activation of the ventral striatum, a subcortical region in the brain which is part of the dopaminergic reward circuit. College students and older adults from the community were recruited for a functional magnetic resonance imaging (fMRI) study which performed brain scans of the participants while they were instructed to work on decision-making tasks. The participants were divided into experimental self-affirmation and control non-affirmation groups. Before the scan was ran, experimental participants ranked a set of personal values, e.g. art, science and religion, in the order of importance while control participants would rank characteristics of a toaster, which they thought would be most important to a typical college student, someone who is not themselves. During the actual scanning procedure, participants viewed paired pictures which depicted high and low ratings of importance and were asked to pick their preference between the two or neither one.

The results showed greater activation of both left and right ventral striatum regions in the self-affirmation than non-affirmation participant groups. There was no difference in ventral striatal activity when non-affirmation participants were choosing between paired high-ranking and low-ranking toaster's attributes, which suggested the reason of lack of self-relevance in the theme of the task. Other regions with increased activation were the medial prefrontal cortex and posterior cingulate cortex, which also play a role in self-processing and evaluation of emotional stimuli (Dutcher et al., 2016; Maddock, Garrett, & Buonocore, 2003). While an explanation of the link between self-affirmed values and positive reward may not be straightforward, this study aids in understanding the difference when the focus is directed towards the self and not others or an inanimate toaster. When it comes to visualizing concepts as broad as the arts and science, we may draw on a rich variety of experiences with us in the role of the lead actor and the university as our performing stage, which is especially relevant to college student participants. And in a wider field of perspectives, the chances that you will stumble upon more instances of positive encounters are greater and they naturally bring about the pleasurable response with them.

This study is the first to link self-positive aspects with the mesolimbic reward region in the brain which is rich in dopaminergic neurons, suggesting an inherent positivity which underlies our self-perception beginning from early childhood. Perhaps, the self is very much attached to reward expectation and a yearning for positive appraisal that help build up one's worth, purpose and meaning in life. There is more to the story of life and humanity than an intricate connection of neurons which transmit electrical signals at nearly the speed of light. To reduce the concept of one's existence down to the mechanistic working of a 3-pound brain is just one in a million pieces of a puzzle. Just as we cannot do without our brain, we can neither live without a constantly beating heart, a protective immune system or a properly working gastrointestinal tract which feeds our brain with the glucose it desperately needs on a second-to-second basis. The point is that the brain is dependent on the rest of the body in the same way that the body is dependent on the brain. The brain and the rest of our body work in concert with each other; they cannot be completely detached from each other in the way we study about them separately in discrete academic disciplines and neither one can exist without the constant support of the other. If you are only your brain, why is cardiac arrest or heart failure a matter of life and death for the emergency patients who flood the hospitals day after day? When a new discipline like neuroscience now comes into the limelight, people simply jump on the bandwagon and start bragging about the miracle working power of the brain, so called the popular phrase "you are only your brain." The brain may be an authoritative control centre that disseminates instructional orders to the rest of the body but a captain without his group of healthy and well-nourished subordinates is merely a captain of a sinking ship.

In as much as dopamine is needed and produced in the brain, there is also a considerable amount of this "multi-functional" neurotransmitter being manufactured in our enteric nervous system, which is responsible for the proper functioning of our gastrointestinal tract. The list could go on to include dopaminergic receptors on our cardiac cells, the lymphocytes of our immune system as well as the kidney cells of our renal system. Such fact stresses the holistic nature of our human body and the implication that certain drugs which affect the processes in the brain could also produce secondary, usually undesirable, outcomes in the peripheral nervous system. Once again, the brain and the peripheral of the body are inseparable entities which constantly rely on each other.

It is interesting that when the theme of reward is being discussed, there is a connotation of a sense of positivity that carries over into the picture. Research has shown that there is a relationship between the widely studied reward learning system of our brain and depression. Typically, depressed individuals exhibited impaired responses to positive or rewarding stimuli which would normally activate the nucleus accumbens, a region in the ventral striatum which is part of the mesolimbic reward processing circuit. Here, dopamine receptors thrive...Although there are other regions which react more intensely to negative stimuli, this observation seems to hint that such striatal region is more sensitive to positive stimuli and demonstrates decrease in activity when exposed to negative stimuli. There could be potentially further exploration in healthy non-depressed individuals through testing whether such differential activation levels could be detected. Thus far, imaging comparisons of brain activity have been conducted for the purpose of contrasting between depressed and non-depressed subjects, which really limit our understanding of wherein specific regions of the brain are more attuned to positive and rewarding stimuli. Following such, there is the question of how to overcome the minimum threshold level required for activation to generate healthy response by depressed subjects when exposed to positive stimuli. On the other hand, there is less elucidation by research interpretations on the underlying connection between emotional response and dopaminergic activation. It appears that the nature and level of intensity of our emotion offer us a dimension of measurement continuum that determines the valence of stimuli and in simple terms, whether they are "good" or "bad." We cannot really begin to tell how detrimental or beneficial a particular stimulus or agent, e.g. a horror movie or a smiling friend, is to us without the consciousness of our emotional response or reaction, which gives weight to the threshold and outcome of our perception. How can you appreciate the smile and friendliness of a person standing right in front of you without your emotion enabling you to feel a certain degree of happiness and positivity towards her? In a sense, how can you be sure that your brain's dopamine levels are not increasing at the same time to facilitate and work in concert with your peripheral nervous system to activate your emotional response? Additionally, the definition of "pleasure" and "mood" can be vague and lacking in specificity in research methodologies. Although it has been reported that dopamine is not associated with pleasure but more so with motivation, the Westernised research definition of pleasure may be more related with fleeting mood experiences and the immediate aftereffect of reward/goal attainment, but yet which is less attached to a global sense of positive appraisal and "stability of happiness" beyond the transient phase of alternating moods in response to reward. The essence and experience of "happiness" is a process that is drawn over the long-term and influences our routine on a daily basis, as opposed to transient uplift and downward plunge of stability lacking "pleasure" and "mood." In light of this, happiness and pleasure/mood are at least dichotomous in terms of measures of duration and (graphical) intensity pattern of effect.

Negativity bias has us looked into the effect of social comparison conflict, pain perception and impulsivity on brain activity, yet the underlying mechanism for positive response and satisfaction, even the tendency to experience happiness as an emotionally stable state as opposed to transient feeling of pleasure, is poorly elucidated. Perhaps, studies are pointing towards the same direction, but we are just not confidently appreciating it. Although dopamine is less strongly linked with the consummatory or pleasure phase of reward processing, just as professors and experts plainly stated that endogenous opioids are more responsible for it, it appears that we could be missing one more (or the third) component of reward processing that may proceed the consummatory phase. On the other hand, one can pose the question of how dopaminergic activation, which begins with the first anticipatory or "wanting" phase, proceeds to involve endogenous opioids in the consummatory phase. It appears that such opioid activation does not work in isolation from dopamine and that the reinforcing effect of mesolimbic reward circuit has been studied to contribute to opioid drug addiction (Meyer & Quenzer, 2013). However, it should be noted that the increased release of dopamine also occurs with many other types of drug addiction, particularly illicit drugs. There is also the question of temporal precedence and bidirectionality involving the sequence of activation between dopamine and endogenous opioids should we want to pass the baton of "pleasure molecule/neurotransmitter" from dopamine to the opioids. It is difficult to cast aside or ignore the emotional component of the anticipatory(wanting) phase in reward processing, which may include some elements of pleasure. On the other hand, male and female study participants may experience such phase differently when emotional aspects, in addition to pure motivation, are to be considered. Perhaps, a third phase of reward processing could be more of an aftermath reflective or appraisal stage which helps put our experience in a better perspective, even strengthening our memories of the first two phases and contributing to a more lasting positive state. Could dopamine levels be raised, and its elevated activity re-established during such phase? There is also the possibility that maintenance of stable and consistent dopamine activity may contribute to a cumulative long-term state of well-being and positive health. It appears that dopamine is a neurotransmitter which plays a major role that influences our personality in a variety of aspects from desire, motivation, goal setting to liking and satisfaction. At least in these aspects of our life pursuits and functioning, the other classical neurotransmitters are less instrumental and impactful.

In the area of cognition and memory, the research literature gives a general impression of the less of contribution by dopamine than the counterparts of cholinergic receptors in the hippocampus, as popularised by the negative impact of Alzheimer's disease. Nonetheless, just as our brain is a deeply connected holistic organ, the dopaminergic neurons from our midbrain region do have an important, if not necessary, function to play for their innervation of the hippocampal region (Gasbarri, Verney, Innocenzi, Campana, & Pacitti, 1994; McNamara et al., 2014). In 2014, a group of researchers at Johns Hopkins and University of California, Irvine, led by Dr. Michael Yassa, tested study participants by administering a surprise memory test a day after they were shown a set of over a hundred pictures and given a pill which contained either caffeine or a placebo. The obvious results were that the caffeine administered group of participants performed better than the placebo group. What is interesting from this study is that caffeine indirectly promotes dopaminergic activation through antagonising adenosine receptors, which are in turn dopamine antagonists. In addition, D1 dopaminergic receptors are also present in the hippocampus and they have been shown to link to our response to novelty (Lemon and Manahan-Vaughan, 2006; Tran et al., 2008). Even in terms of exposure to a novel spatial environment using experimental rat subjects, the level and nature of activation of glutamatergic neurons in the hippocampus, which promote memory storage, are dependent on the activity of D1/D5 dopaminergic receptors (Li, Cullen, Anwyl, & Rowan, 2003; Menezes et al., 2015). Here, we see a co-dependency of more than one class of classical neurotransmitters working in concert, almost in a synergistic fashion that lends an example of a situation in which no single neurotransmitter is solely responsible for one kind of brain function or faculty. United and cooperative neurotransmission activity work best! Thus, novelty response and retention are significantly aided by dopaminergic activity which facilitates further memory formation and storage. Perhaps, this offers a clue why first-time impressions and memories remain with us indefinitely, at least for healthy individuals. Furthermore, the emotional and arousal components of novel memories are, in the intuitive sense, richly integrated by the hippocampal connections with the mesolimbic reward pathway (Duszkiewicz, McNamara, Takeuchi, & Genzel, 2019; Wittmann, Bunzeck, Dolan, & Düzel, 2007). And, one cannot deny that positive novel experiences often come packed with pleasurable feeling and meaningful content. Thanks to dopamine??

In terms of physical health, dopamine also comes into play beyond the boundaries of our brain and nervous systems. A recent animal model research project also talks about the link between the dopaminergic reward system in the brain and the immune system's ability to control tumour growth in the animal subjects. Interestingly, dopaminergic receptors are also expressed on important immune cells in the body - T- and B-lymphocytes, which play critical roles in immunotherapy. Dopamine also has an anti-inflammatory role which is still under-explored. In one mice model study, the activation of dopamine D2 receptors by agonists suppressed neuroinflammation and ameliorated brain edema (Zhang et al., 2015). It seems that dopamine, apart from being the most popular classical neurotransmitter, is also increasingly being recognised as an immuno-modulator. It may even have the potential to work as an anti-inflammatory agent. On the other hand, dopaminergic receptors are also present in the heart and the treatment of patients with heart failure with low dose dopamine helps promote diuresis and maintain renal health (Xing, Hu, Jiang, Ma, & Tang, 2016).

In a sense, happiness is more than mere transient feeling of pleasure, attainment of reward and life satisfaction, but a stable and lasting state of mind, emotion and positive energy level. Happiness is therefore closely tied with our perception, both as a sensory and response mode, and the attitude of resilience during times of emotional and physical challenges. Thus, how far can dopamine take us in our quest for positivity in life, only time will tell...

Chapter 2: Blood Pressure Regulation

Research in the medical sciences has explored the functions and applications of dopamine in a number of our physiological systems, such as that of cardiovascular, immunity, digestive and renal. The metabotropic class of dopaminergic receptors which consists of five subtypes (D1, D2, D3, D4, and D5) can all be found in the kidney and thus play important roles in the regulation of blood pressure through excretion and retention of sodium chloride by this organ. Studies of genetic mutations of the relevant receptor subtypes tell us that their dysregulation can impact heritable hypertension in such subgroup of patients in the population. While impaired or inefficient excretion of sodium chloride by the action of mutated dopaminergic receptors in the kidney is associated with blood pressure increase, there are also studies which show that these receptors also interact peripherally with the renin-angiotensin system in the process of blood pressure regulation and still others that imply their vasoconstricting effect as the main contributor to certain cases of hypertension. What is just as interesting is the observation that the action of dopamine on the peripheral renal system in the case of hypertension seems to work in isolation with the brain's cerebral system. All these empirical evidences point to the versatility and diversity of an important classical neurotransmitter that has wide-ranging impact in our overall physiology, although the same can be said for other neurotransmitters.

It should not be taken lightly of the possibility that the regulation of blood pressure by dopamine could have concurrent and lasting impact on other closely related cardiovascular and immune systems. Overall, the effect of dopamine on blood pressure may be at best modulatory, depending on the subtype(s) of receptor(s) at work and the systems it is interacting with and influencing, not to mention the functional outcome it is purposed to produce. In the case where physical exertion is necessary, dopamine may work with cortisol production to enhance blood pressure for increased energy expenditure during the process of the activity, affecting heart rate and respiration as demanded to meet the individual's willful goals and determined tasks. Perhaps, this could hint at the reason why dopamine inhibits motility in the upper gut of our gastrointestinal tract - to spare and channel energy resources away from digestive activity for other immediate physical needs. One may ask, would this impact our vascular and immune system in the long run? Well, one study suggests that exercise-induced increased levels of cortisol may reduce the production of helper-and cytotoxic- T cells, though any mediating role of dopamine in association with blood cortisol has yet to be determined. On the other hand, dopamine receptors are also expressed on T cells as well as B-lymphocytes and play a modulatory role on them. In other words, the action can be enhancing or suppressing (or both) to maintain an optimum level for normal efficient functioning. Such is the fascinating complexity of the human body...Once more, we are only beginning to scratch the surface of it all...The next frontier of neuroscience may just stretch beyond our cerebral system to down the winding paths of our peripheral nervous system...

The dopamine theory of schizophrenia should be very famous by now. Experts seem to be attributing the "full" cause of psychotic symptoms to excessive dopaminergic activity in the brain and could not find better alternative candidate(s). Drilling down further, this overactivity also puts the blame on the D2 receptors in the mesolimbic area and less on other subtypes of dopamine receptors. Is this supposed to be the whole picture? As expected, inconsistencies which challenge this theory are surfacing and it appears that the empirical foundation which has only loosely supported the causal relationship between D2 receptors and psychosis is being shaken. Since D2 receptors produce an inhibitory response as opposed to excitatory post-synaptic potentials (EPSPs), the activation of these receptors is going to decrease electrochemical activity in the post-synaptic cell. If this fact is unquestionable, then the association of such inhibition with a contradictory "overactivation" of dopaminergic transmission is more questionable. Moreover, since D1 subtype receptors are excitatory, there is the probability that D1 antagonists will produce the same outcome as D2 agonists. How about the opposite - D2 antagonists (antipsychotic medications) producing similar effect(excitatory) as D1 agonists? It would be interesting to discuss these issues in-depth. When it is known that there are many more D1 receptors in the prefrontal cortex area, research tends to go straight towards them and takes lightly about the D1 receptors in the ventral striatum. Although D1 receptors are not as abundant in the ventral striatum as D2 receptors, their activation might somehow compensate for the excessive inhibitory effect of D2 receptors. More research on this subject is of necessity.

Visual Broadcasting: A Probable Extension of Schizophrenia Symptomatology

A previously atypical symptom report of a hospitalised patient experiencing unusual visual broadcasting has been documented but has received minimal attention in the research and clinical field of psychiatry. A preliminary hypothesis is proposed of an implicit mechanism of highly integrated contra-directional neural processing of the visual sensory and cerebral perceptual faculties that might have been triggered by a global heightened activation of the dopaminergic system which leads to a concurrent "visual and thought" psychotic episode. A follow-up on the insight and comprehensive understanding of such atypical and novel symptom as an extension of schizophrenia is warranted.

It has been a pleasure to read with interest the letter article entitled "Visual Broadcast in Schizophrenia" from the reputed journal of BMJ Medical Humanities, which touches on a novel and rare symptom manifestation of schizophrenia that accompanies persistent paranoia, imagined persecution and thought broadcasting of a patient under the care of the authors (Hunter, Mysorekar, & Woodruff, 2005). To date, repeated searches in the literature regarding this unique and patient-specific symptom have been performed, yet without yielding new result that is beyond the publication of the aforementioned article. However, this should not deter the research community from gaining insight and deeper understanding into this sensory broadcasting symptom that is potentially detrimental on the health and wellbeing of severely ill psychiatric patients, yet likely escapes documentation. As a preliminary attempt to explain this symptom occurrence, it helps to analyse from a perspective of an exacerbation of a cluster of multiple interrelated symptoms that preceded an extrasensory modality involving the patient's visual sense and perception. Since this symptom is not experienced by a majority of the patient population but by an admitted inpatient, it is suggested that the symptom of "visual broadcasting" which expresses as a modified form of thought broadcasting in which visual images which can be physically seen in reality by the stated patient were being broadcasted to the minds of others, likely manifested as an outcome of extreme stress circumstances. Intuitively, there could be a concurrent upsurge in dopamine release as an opposing response to high intensity emotional and mental distress denoted by multiple symptoms of the patient. It can be argued from the point of view of the concept of stress-induced psychosis that the more cortisol is produced in the body during intense stress, the more dopamine is released in excess to counter or oppose the ill effects of cortisol, based on evidence of positive correlational relationship between cortisol and dopamine release in human subjects (Mizrahi et al., 2012; Wand et al., 2007). On the other hand, the potential emotional relieving benefit of dopamine in response to stress-induced cortisol increase could be implied from a study by Hamidovic and colleagues (2010) who reported positive correlation between elevated cortisol following performance of a stressful task by healthy subjects and positive mood induced by administration of amphetamine. As amphetamine has been known to increase dopamine concentrations in a number of studies, this correlation points to an intimate relationship between cortisol and dopamine that is both complementary and counteracting which could potentially affect vulnerable healthy human individuals (Carboni, Imperato, Perezzani, & Di Chiara, 1989; Leyton et al., 2002; Pontieri, Tanda, & Di Chiara, 1995; Wyvell & Berridge, 2000). Such correlational release may be functional in the manner that there is an underlying purpose to counter the negative valenced emotional outcome of increased cortisol through the promotion of action-oriented motivational response by dopamine. Rothschild and colleagues (1984) once reported in their study that the levels of plasma free dopamine in their unmedicated human subjects were considerably approximately 3-4 times higher after two administrations of a small dose dexamethasone than immediately before administrations in the morning and evening, offering hints of support to their explanation that corticosteroids such as dexamethasone may increase dopamine and subsequently produce psychotic experiences in unmedicated individuals. Although the underlying association between increased plasma dopamine and its corresponding release in the brain is open to investigation, the finding of depressed individuals with higher cortisol levels being prone to psychosis warrants consideration of its significance in a deeper understanding of the etiology of schizophrenia. Perhaps, cortisol and dopamine have mutual functional significance which is intimately linked in the event of acute stress to provide emotional and physiological benefits for an individual. Therefore, it does not come as a surprise that antipsychotics have an effect on both cortisol and dopamine levels in the body, i.e., there were reduction in salivary and serum cortisol levels after treatment (Mondelli et al., 2010; Venkatasubramanian et al., 2010). The condition of stress-induced psychosis de-emphasises geneticism as a sole reasoning basis for the causation and stability of traditional prevalence rates of schizophrenia in societies (Foley, Corvin, & Nakagome, 2017; Henriksen, Nordgaard, & Jansson, 2017). We should ask the question why psychosis tends to first manifest its first episode during late teens and early twenties in the youth population. During such a critical period of transition in a young individual's life, peer pressure, high school to university environmental and performance demands and complex relationship changes all combined to induce heavy stress on a highly constrained cerebral space of the brain which may well still be developing onward to full maturity. It gives even more reason to say that this particular sensitive age window during one's youth in life leaves wide open an individual's vulnerability to psychotic manifestation. Therefore, heavy educational and social demands placed on a sensitive brain during such a burdensome transitional period in life have a significant role to play in the etiology and development of first psychotic episode in youth. In addition, it also helps shed light on an alternative target of intervention for stress-induced psychosis, i.e., cortisol other than dopamine D2 receptors, through cognitive and emotional stress regulation channel. No doubt, elevated cortisol and corticotropin releasing factor (CRF) levels affect other aspects of normal physiological activities such as sleep, therefore their increase in chronic insomnia patients, due to cortisol's functional connection with a heightened sympathetic nervous response that subsequently induces greater vigilance and consciousness arousal (Backhaus et al., 2006; Chen et al., 2016; Rodenbeck, Huether, Rüther, & Hajak, 2002; Roth, Roehrs, & Pies, 2007). Such close relationship and the knowledge that sleep adequacy or disturbance has an important role to play in the onset of psychosis and hypervigilance (and may be more pronounced during the crucial stage of adolescence due to academic and social pressures) further sheds light on the complex web of causality and effect of schizophrenia. (Question for Thought: Does a person who find himself not sleepy at night represent a sign of cortisol overactivity?)

There could be latent motivational consequences of dopamine upsurge during psychosis in response to stress induction, and it should be noted that one of the negative effects towards health of increased level of cortisol is tissue inflammation. There have been indications that dopamine could potentially lower inflammation based on its varied immunomodulatory functions on a short-term basis (Beck et al., 2004). Although proinflammatory markers have been shown to be elevated in schizophrenic individuals, it is still questionable whether there is a direct causality link between excessive dopaminergic activity and inflammation in the brain, as opposed to invasive pathogenic substances (Müller, Weidinger, Leitner, & Schwarz, 2015).

Further investigation into this hypothesised function of excessive dopamine release in response to stress induction could shed light into the purposeful etiology of psychosis in vulnerable individuals, thereby introducing a novel understanding of not just how, but why schizophrenia came to be. In consideration of the above discussion linking cortisol and dopamine release, one might wonder about why the choice of these two hormone and neurotransmitter. Let us refer back to chapter 2 on the important modulatory role which dopamine and its receptors play on our blood pressure regulation. Each dopamine receptor subtype is expressed on the organ components of our kidneys and renal system. In addition, there is a close relationship between the regulatory processes of our blood pressure and the sympathetic nervous system. These seemingly discrete physiological processes and subject matter are very much interlinked like a harmonious orchestra due to the underlying involvement of dopamine. The two classes of receptor subtypes, i.e. D1-like and D2-like, are both opposing and complementary at times, depending on the immediate and long-term needs of and demands on our body. This is analogous to the varying degree of distinct expressions of D1-like and D2-like receptors in our prefrontal cortex and midbrain region respectively, which is why psychosis is mostly attributed to the overactivation of the D2 receptors in the mesolimbic reward pathway. Similarly, D1-like receptor agonists in the renal system have been associated with decrease in blood pressure and it is the impairment of D1-like receptors that is linked to genetic hypertension through their reduced ability to decrease sodium transport and reabsorption (Jose, Eisner, & Felder, 2002). On the other hand, D2-like receptors seemed to have an opposite function as well as being a role player in vasoconstriction of vessels during activation of sympathetic responses, just like what an increased release of cortisol does to the body. However, it should not be forgotten that there is a constant synergistic and mutual interaction between the individual D1-like and D2-like renal receptors working in wave-like patterns on a graph to maintain both functions and stability for our physiological benefits. Intuitively, we cannot ignore the role of the amount of energy resource that is being supplied to the brain at any one time. When blood and sugar are focused on being delivered to the mesolimbic region, say, during a psychotic episode, physiological constraints and limitations may naturally redirect a normal supply of these nutrients away from the D1-receptor-rich region of the prefrontal cortex for the sake of a greater proportion devoted to the midbrain, hence a reduced ability of rational cognitive reasoning effort that can be afforded at such instant. As you may have guessed by now, D2-like receptors are more closely associated with hypertension through increased vasoconstricting function on the sympathetic nervous system.

With an understanding of the functional role of dopamine during psychosis in mind, current research also lacks an exploration of a likely controversial yet potential range of neuroprotective benefits of stress-induced psychosis that leads to surge of dopamine release on a short-term basis. One study reported dopamine-infused reduction and increase in intracranial pressure in patients with cerebrovascular diseases, which brings us to the question of an underexplored relationship between emotional stress and intracranial pressure (Nau, Sander, Klingelhöfer, 1992). On the other hand, Ract and colleagues (2001) reported that infusions of dopamine on a model of brain trauma which raised intracranial pressure did not increase cerebral perfusion pressure. Cerebral perfusion pressure helps manage blood flow to the brain, and as increased intracranial pressure limits such flow, decreased cerebral perfusion pressure might indirectly hinder an overactive state of neurotransmission. In addition, Myburgh and colleagues (1998) showed dopamine significantly increases intracranial pressure in animal subjects. In their review study, Morrison, Frame and Larkin (2003) found evidence which supports trauma as having causality relationship with psychosis. This points to a suggestion of a broader connection between psychological and physical trauma that might share modest similar structural alterations and impact on the vulnerable brain. Future investigation should attempt to elucidate a potential correlation between psychological distress and intracranial pressure, an intersectional frontier between psychology and medicine which the current state of research has greatly undermined, due to the underlying Western philosophy that emotion in psychology is distinctive from physiology.

The precedence of pre-existing paranoia and thought disturbances in the aforementioned patient, presumably induced by excessive dopaminergic activity, probably has spread to the sensory and perceptual domains which potentially involve a specific population of D1/ D2 receptors in the retinal cells proceeding internally to the brain to warrant a distinct category of paranoia symptoms (Brandies & Yehuda, 2008; Tian, Xu, & Wang, 2015). Research has discovered the presence of these D1/ D2 receptors in the retinal cells, which brings us to question the extent to which dopaminergic receptors beyond the mesolimbic reward pathways could present vulnerability to psychotic disturbances due to structural and functional similarities shared between same classes of receptors as one likely plausible explanation for the "visual broadcasting" symptom. Reports of low cortical signal-to-noise ratio in schizophrenia simply indicates that there is extensive background noise neurotransmission occurring within cortical networks which lends to a form of activity-caused burden on intracerebral space (Winterer G, Weinberger, 2004). Whether this might lead to a form of outward-directed "diffusion" or "dispersion" of wave-like electrical activity in reality(as opposed to the widely accepted belief in psychiatry that such symptoms are imaginary) in order to ease the distressing burden on intracranial space limitation remains to be further investigated. On the other hand, there is also the question of how the retinal cells connect with the population of receptors in the visual association cortex to affect one's perceptual and cognitive distortions.

The above suggested mechanism of connective pathway(s) between visual sensory and perceptual cognition hints at a highly integrated and connected neural system consolidating both central and peripheral networks, i.e., visual sensory and cerebral faculties of perception and cognition, which is often neglected in the discrete learning and disciplinary approach in Western traditional medicine. In addition, there could be an underlying latent synchronization of dopaminergic activity at play on a global scale as opposed to concentrating on localised units, rendering a simultaneous manifestation of "visual and thought" psychotic episode. Intuitively, if such distinct symptom can be categorised as sensory/perceptual distortion, the neural processing involved could be undergoing a reversal in direction that instead of the perceived visual properties of stimuli being directed outside-in, it ended up inside-out, in a contra- or opposing pattern that has surpassed the boundary of normality.

The patient documented in the report by Hunter et al. (2005) above experienced this severe grade novel symptom of visual broadcasting, but this should not preclude the likelihood of an added auditory component that is beyond the common hallucinations prevalent in schizophrenia, i.e., there could be real audible sounds which are " diffused outwardly and transmitted" to others. Coincidentally, dopaminergic inputs to the auditory cortex are also present in the brain and they include the expression of D2 receptors which aid in sound discrimination learning that can be incentivised by reward in animal studies (Kudoh & Shibuki, 2006). Such observation should encourage the opportunity of reassessing the potential sensory role of dopamine beyond reward, learning and memory, yet also the overlooked role of D2 receptors in learning in the nervous system, as opposed to exclusively of reward pathways and psychosis. . Although research on dopaminergic involvement is mostly concentrated on the inner brain's mesolimbic reward pathways, the actual mechanism of action of D2 type receptors, which primarily contribute to psychotic disturbances, cannot be ruled out on peripheral populations of D2 receptors elsewhere that bear similar neurochemical and structural makeup. It would certainly be beneficial if the authors could share further detail on the course of medication which has helped alleviate this symptom of the patient in the article, thus potentially offering positive hope for others who could have been affected but are suffering in silence and isolation. In addition, an interactive process of comfort and assurance by the attending psychiatrist that is patient-centred and focused could potentially hasten recovery if utilised to a greater degree than currently in place in most institutions.

On the other hand, it is a wonder whether D2 receptors, as opposed to other subtypes, are more involved in reward (both appetitive and consummatory stages) and addiction processes. If this is the case, shall we redirect our focus on targeting D2 receptors, which are widely distributed in the mesolimbic reward pathway, for substance abuse/addiction disorders? While dopamine antagonists are very useful as antipsychotics, it would be unrealistic to assume that their long-term use does not come with undesirable effects.

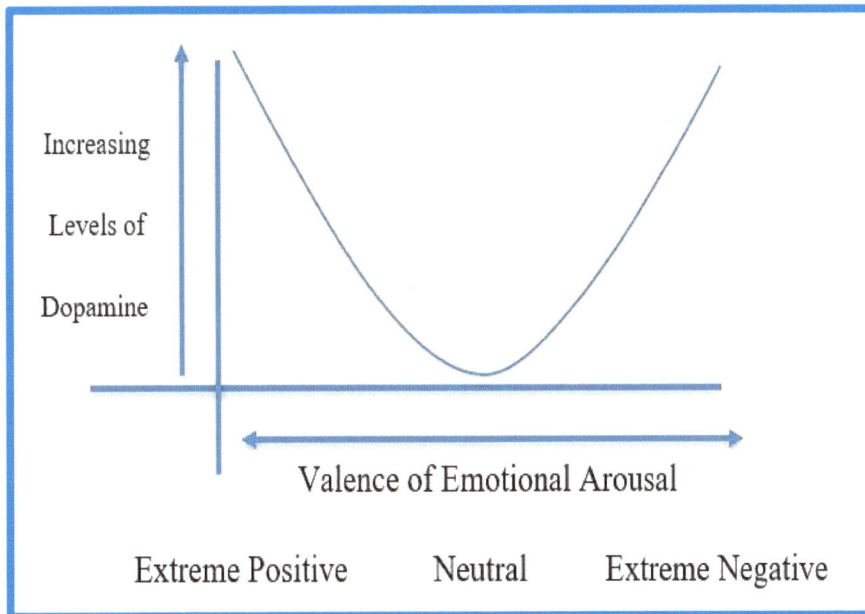

Figure 1: Hypothesised relationship between dopamine levels and valence of emotional arousal based on evidences of psychosis and activation of reward pathway.

Antipsychotics, or neuroleptics as they were once called, are medications used to treat psychotic symptoms, e.g. hallucination and delusion which are commonly associated with the psychiatric disorder of schizophrenia. Although patients and physicians are readily familiar with their most common side effects of metabolic risks of weight gain and raised lipid and cholesterol profiles, many may not relate such class of hormone level-altering drugs with the risk of brain neuropathology. Nevertheless, there is a link between the brain neurotransmitter dopamine, the activity of which is controlled by anti-psychotics when a person is taking them, and a hormone called prolactin which is produced by the pituitary gland or "master gland" in the brain that secretes hormones and therefore controls the endocrine system of our entire body.

Prolactin is a hormone which has the main function of inducing lactation or stimulation of milk production from the breasts of females after childbirth. However, this is not the only function of prolactin for it also helps to regulate a variety of processes of the repro-ductive and immune systems, just to name a few. Dopamine, in being produced by a nearby region called hypothalamus which oversees the functions of the pituitary gland, helps regulate the production of prolactin by decreasing it and through a negative feed-back loop. However, with antipsychotics, which reduce the activity of dopamine, a condition called hyperprolactinemia, or increased production of prolactin, can develop.

Although a clarity of the mechanism responsible for the association has yet to be investi-gated, the research literature has reported cases of administration of antipsychotics with the growth of a type of pituitary tumor or prolactinoma, a typical pathological cause for hyperprolactinemia. However, it is plausible that with the lack of restraint against production of prolactin by dopamine, the action of antipsychotics which dampens the neurotransmitter's activity on receptors may have eased the process of growth of prolactinomas. Interestingly, in a case report of a woman treated with amisulpride, a highly selective antipsychotic for dopamine D2 and D3 receptors, it is suggested that pituitary adenoma or tumor, along with an elevated prolactin level, has been induced by the medication post-treatment (Perroud & Huguelet, 2004). Following this report, an additional three cases of both male and female schizophrenic patients who demonstrated elevated levels of prolactin due to treatment with amisulpride have also been reported (Akkaya et al., 2008). In this group of patients, a post-treatment magnetic resonance imaging scan of the brain after 6 months showed the presence of pituitary microadenomas, although it has yet to be confirmed whether they are prolactinomas and that the possibility they have developed the tumors prior to treatment cannot be ruled out. With such limitations, the relationship can only be regarded as correlational at best, but it does hint at a possible role of an antipsychotic medication on a long-term basis.

On the other hand, a retrospective study of several types of antipsychotics concluded that the medication risperidone, a potent dopamine D2 receptor antagonist, may be associated with pituitary tumors upon analyses of drug adverse events, which was also supported by results of animal studies (Szarfman, Tonning, Levine, & Doraiswamy, 2012). Moreover, the extent of tumor-related adverse events happened to correspond with the level of effect each type of medication has on dopamine receptors, with the highest association being risperidone. This medication has also been shown to demonstrate consistency with the greatest elevated levels of prolactin (Akkaya et al., 2008). Along with a mice model study which induced the removal of prolactin and dopamine receptors, research shows that the activation of dopamine D2 receptors and their opposing or inhibitory effect on prolactin secretion and the fast reproduction of prolactin-producing cells in the pituitary is necessary for the prevention of dopamine-related induction of hyperprolactinemia and prolactinomas, which were more pronounced in female than male mice models (Schuff et al., 2002).

With the development of new generation antipsychotics which act less strongly on dopamine receptors, e.g. aripiprazole and brexpiprazole, the risk of such adverse tumor development could have been minimized and less frequently reported in the research literature as opposed to a decade ago. However, it is wondered how effective such new generation of medications will be for patients with a high degree of psychotic symptom severity. Patients should therefore actively engage in discussion with their physicians upon noticing signs of milk discharge from their breasts and discontinued menstruation as a likelihood of hyperprolactinemia. As a woman's elevated prolactin level may cause her periods to stop, such effect may also increase the risk of infertility. It is therefore necessary that the costs and benefits of medication options be assessed with the physician and an optimal treatment be decided upon. Furthermore, there was a report which highlighted a case of low bone mineral density associated with prolactinoma and high prolactin level in a young male patient (Sperling & Bhatt, 2016). While this patient is otherwise healthy and non-psychotic, it nevertheless emphasizes the importance of maintaining prolactin levels within normal range to reduce the risk of development of osteoporosis (a bone tissue condition which increases one's vulnerability to fractures) and potential complications with related physiological processes which involved increased prolactin levels.

There may be a wonder whether researchers could design alternatives to D2 receptor antagonists/partial agonists for a novel way to treat schizophrenia. Perhaps, we should explore a different approach and target another mechanism. In Europe and elsewhere outside North America, people have been experimenting with selective serotonin reuptake enhancers (SSREs), e.g. Tianeptine, which function in the opposite manner to selective serotonin reuptake inhibitors (SSRIs) (Mennini, Mocaer, & Garattini, 1987). These reuptake enhancers increase the ability of certain neurotransmitter transporters in the presynaptic neuron in taking back up the excess serotonin that has been previously released into the extracellular synaptic cleft. Such reuptake enhancers are rare and have seldom been researched compared with reuptake inhibitors. It could be that substances which possess such function are very difficult to discover and obtain. However, we could be positive that they are not inexistent in the natural world. Zhang and colleagues (2010) experimented with a natural flavonoid substance called luteolin and its derivatives and discovered them to be potent dopamine transporter agonists. However, there has been little to no follow-up work performed on this under-appreciated preliminary investigation since then, in particular along the line of its potential clinical application in antipsychotic treatment. If there just happens to be more candidates in the queue which qualify as dopamine reuptake enhancers as they are for serotonin, perhaps this could be a new pharmacological target for the treatment of schizophrenia, hopefully bypassing the Parkinsonian and extrapyramidal symptoms, tardive dyskinesia, hyperprolactinemia, weight gain and a number of other side effects associated with typical and atypical antipsychotics. Furthermore, there is the potential that reuptake enhancers, as present mostly on presynaptic neurons, could be more selective than current antipsychotics, which affect more widely distributed D2 receptors that, as we have seen, are expressed in other organs and systems outside the brain as well. It could be only a matter of time when these reuptake enhancers will be falling within our reach…

Chapter 4: Protective Strategy for Dopaminergic Neurons in Parkinson's Disease

They are too common in our everyday life and language to be ignored. When physical symptoms call for our action, we reach for the bottle of acetaminophen for pain relief, whereas to satisfy our craving, we aim for the kitchen coffee-brewer or the nearest cafe. They are *effective* and have always been so. But are these remedial actions just about all or the best that acetaminophen and caffeine can offer us? It seems that scientific research is beginning to paint the fuller picture. Experts are tapping into the drugs' less obvious but potential useful functions that could be relevant in the treatment of Parkinson's disease, a neurodegenerative disorder which debilitates an individual's motor and muscular ability. While the causes of Parkinson's disease remain unknown, genetic factors do not appear to contribute to a majority of the cases and epidemiologic studies have also focused on studying health and lifestyle factors and their correlations with the risk of disease development (Carlson, 2011; Simola, Pinna, Frau, & Morelli, 2014).

Studies which reported that participants' caffeine intake lowered one's risk of developing Parkinson's disease led to promising proposed theories of mechanisms of how adenosine receptor A_{2A} antagonists provide neuroprotective benefits by modulating the excitotoxicity of glutamate-induced activity in the subthalamic nucleus (Simola et al., 2014). They also indirectly attenuate neuroinflammation through action on microglia and astrocytes, cells which provide protective and supportive functions to neurons. It has been known that caffeine renders us more alert by blocking adenosine receptors, A_1 and A_{2A}, as the level of adenosine increases with periods of wakefulness and helps promote sleepiness (Carlson, 2011; Kong et al., 2002). In the case of Parkinson's disease, the focus is on the multiple negative effects of stimulated A_{2A} receptors on nigrostriatal dopaminergic system, the pathway involved in movement control, which connects the substantia nigra with the striatum and a target region of treatment against the disease. The degeneration of dopaminergic neurons in this region contributes to the motor deficit symptoms of Parkinson's disease and is thought to increase glutamatergic input from the cortex to the striatum. This in turn elevates glutamatergic activity in the subthalamic nucleus and results in excitotoxicity effect on neurons in the substantia nigra, which are very close by in the region. Studies have demonstrated that the death of dopaminergic neurons in the subthalamic nucleus and decrease in dopamine levels in the striatum can be counteracted by A_{2A} antagonists as the stimulation of such receptors increased the level of extracellular glutamate (Greenamyre, 2001; Lancelot & Beal, 1998; Morelli et al., 2010; Popoli, Betto, Reggio, & Ricciarello, 1995; Schwarzschild et al., 2003). From this proposed theory, we get to have a glimpse of the intimate interconnection between different brain regions which function together to affect our neurophysiological well-being.

As mentioned, second mechanism of A_{2A} antagonists works by countering neuroinflammation through suppressing the activation of microglia and astrocytes, which also express

A$_{2A}$ receptors and initiate inflammatory responses (Armentero et al., 2011; Halliday and Stevens, 2011; Hirsch and Hunot, 2009; Litteljohn, Mangano, Clarke, Bobyn, Moloney, & Hayley, 2010; Lopes, Sebastião, & Ribeiro, 2011; Reale et al., 2009). Neuroinflammation plays a role in the neurodegeneration progress of Parkinson's disease. It is worth a note that the action of of A$_1$ receptor antagonists did not produce comparable results of neuroprotective benefits as those of A$_{2A}$ receptors in mice models of Parkinson's disease (Chen et al., 2001).

As multiple processes are involved in the progression of Parkinson's disease, there are treatment options which may target certain mechanisms but at the same time, not impacting others. Therefore, a single drug or surgical procedure may not suffice to keep the full range of symptoms at bay. The degeneration of dopaminergic neurons, apart from its initiation by genetic mutation and protein misfolding and aggregation which are followed by a series of cellular disarray that culminates in motor dysfunction of an individual, is also exacerbated by oxidative stress. Research has looked into the potential role of the over-the-counter pain remedy, acetaminophen in reducing the effect of such process on disease progression. In an animal model study, administration of low concentrations of acetaminophen have been shown to be protective against neurodegeneration induced by 6-hydroxydopamine(6-OHDA), a neurotoxin which can self-oxidize and generate reactive oxygen species to deplete anti-oxidant enzymes within cells and this leads to eventual cell damage (Locke, Fox, Caldwell, & Caldwell, 2008; Simola, Morelli, & Carta, 2007). In addition, various concentrations of acetaminophen in the same study were effective in suppressing tyrosine hydroxylase(TH)-induced degeneration, also a contributor to oxygen radical formation that harms dopaminergic neurons (Adams Jr., 2012; Locke et al., 2008).

A study by Tripathy and Grammas (2009) which tested the response of rat brain endothelial cells, that were pre-treated with acetaminophen, to oxidative stress demonstrated increased cellular survival when exposed to menadione, the stressor which released reactive oxygen species. The protection offered to these cells was also due in part to the ability of acetaminophen to increase the expression of an anti-apoptotic protein Bcl2 and thus negatively influenced cell death. A later study reported that the addition of acetaminophen to the brain microvessels of rats increased vascular expression of neuroprotective proteins, which were discovered to decrease with increasing age of the animals (Tripathy, Sanchez, Yin, Martinez, & Grammas, 2012). Although further research is needed, and not just animal models but also human epidemiologic studies of acetaminophen consumption and in spite of the drug's mechanisms being less direct and target-specific compared with caffeine or A$_{2A}$ antagonists, such preliminary evidence should not be taken lightly just because they sound trivial and too good to be true.

While the etiology of Parkinson's disease may be as elusive as the discovery of its ultimate cure, we could glean useful information from the established risk factors, e.g. age and gender (Van Den Eeden et al., 2003). It is widely known that old age and being male increase one's risk of developing Parkinson's disease. Based on such observation, research has looked into lifestyle factors commonly associated with the male gender, from smoking habits to alcoholism (Benedetti et al., 2000; Chekoway et al., 2002). The surprising outcome is that these two popular health-compromising lifestyle habits could be beneficial instead of doing more harm. On the other hand, it is interesting to note that in a part of the nigrostriatal pathway brain region where dopaminergic neurons tend to degenerate in Parkinson's disease, the substantia nigra pars compacta, there are differences in the expression of androgen and estrogen receptors present in the cells through the stages of development which may regulate other neurotransmitter systems (Ravizza, Galanopoulou, Velišková, & Moshé, 2002). Studies have also shown that estrogen may have neuroprotective functions for these neurons despite its antidopaminergic action (Morale et al., 2006; Morissette, Al Sweidi, Callier, & Di Paolo, 2008). Future research should delve into the role of testosterone in influencing the maintenance of dopaminergic neurons of this sensitive region in relation with Parkinson's disease.

Chapter 5: Dopamine, Dysfunctional Reward Processing, and Depression

Dopamine is known to be involved in many brain functions, particularly motivation, learning, reward and pleasure processing and motor control, with research beginning to discover more of its role in depression and related neuropsychiatric disorders. Depression is increasingly being recognized as a multi-faceted disorder which not only affects one's mood and affect, but also daily functioning, cognition and motivation, with the aspect of impaired reward learning playing a role in the manifestations of its symptoms. To achieve a comprehensive account of the etiology of this disorder therefore requires a consideration of multiple factors which renders depression to be one of the most disabling and counterproductive to individuals with such condition. A growing number of research has focused on studying the dysfunction of reward processing pathways exhibited by depressed individuals, however drug intervention which targets dopaminergic regions is still in its early stages of experimentation with less specific direction to take. Furthermore, it is unclear as to what early factors contribute to the impaired reward processing and imbalance in sensitivity to positive and negative rewards displayed by individuals with depression. Inconsistent and ineffective drug treatment outcomes for clinically depressed individuals have encouraged further research of the disorder in various directions, such as cognitive-behavioral strategies and brain stimulation, with the goal of expanding the range of better and promising alternative therapies in alleviating persistent symptoms. Previous understanding which strongly associates neurotransmitter serotonin imbalance with depressive disorder bears limitation in explaining causality and present research now turns to the dysfunctional reward processing perspective (Delgado, Charney, Price, Landis, & Heninger, 1989).

The mesolimbic dopaminergic reward pathway of the brain which is made up of neuronal projections that run from the midbrain region to the nucleus accumbens has long fascinated researchers with its significant role in reward-seeking behaviour, addiction and impulsivity. Research is increasingly exploring the theory of impaired reward processing which could contribute to the psychopathology of depression. A reward is anything with an attractive value which creates a tendency for an individual to approach and consume it (Schultz, 2015). These approach and consuming behaviour are the anticipatory and consummatory components of reward processing respectively. Studies with depressive patients have demonstrated their reduced sensitivity to reinforcement of positively rewarding stimuli, possibly due to dysregulation in reward learning which happened to also lead to atypical response of decreased aversion of negative stimuli (McFarland & Klein, 2009; Murphy, Robbins, & Sahakian, 2003). Anhedonia, a decreased response and affect towards pleasurable stimuli, is exhibited by depressive individuals. Pizzagalli et al. (2009) reported that a region in the basal ganglia, the caudate, has reduced volumes in depressed study participants which were associated with their higher scores of anhedonic symptom. This stresses the role of volumetric brain differences between depressed and non-depressed individuals that could account for the reduced interneuronal connectivity and sites of activation in regions where dopaminergic receptors are concentrated. To achieve a more thorough understanding of the psychopathology of depression, it is important to draw relevant connections between the social basis of reward-related learning and the accompanying underlying neurochemical processing in the brain's internal network which together promote the vulnerability to the disorder. Thus, early life experiences and brain development and neuronal organization during such crucial stages of childhood and adolescence are significant in their contribution to the shaping of one's attitude towards reward-related learning and the behavioral tendencies, thus the extent of insensitivity towards value of reward and the severity of dysfunctionality which result and act on the neurochemical processes of brain development. Nevertheless, there is also bidirectionality in the way one interacts with the environment and the genetic susceptibility to certain types of stressors unique to an individual should not be excluded from the overall picture.

Increased Focus on Dopamine and the Reward Circuit

Research which has received much recent attention also delves into the role of dopaminergic reward system in the neuropsychopathology of depression, paying particular attention to impaired response to positive rewarding stimuli in depressed individuals (Dunlop & Nemeroff, 2007; Epstein et al., 2006; Nestler, & Carlezon, 2006; Pizzagalli, 2009; Ruhé, Mason, & Schene, 2007; Tremblay, 2005). Dopamine is a neurotransmitter which belongs to the class of monoamines like serotonin and is known to be involved in a variety of mental and physical functions such as motivation, reward learning and locomotor activity (Beninger, 1983; Flagel, 2011; Wise, 2004). The implication of decreased activation of dopaminergic neurons being involved in the mechanism of depression is that the progression of the disorder would have a far-reaching impact on a number of aspects of an individual's functioning as opposed to solely mood dysregulation. The activation of dopaminergic neurons for processes required in motivation, learning and movement which leads to performance of action by an individual can have a role to play in the depressed physical functioning of the disorder. Although it remains to be tested, the negative effect of impaired reward reinforcement learning in depression likely develops over time to produce lasting effect into adulthood. Exploring the variation between short-term and long-term depressive periods of study subjects in relation to imaging evidence of dysfunctional dopaminergic activity could offer insight into the temporal workings and probable causality role of the brain's reward circuits. We have seen in a previous chapter that reduced dopaminergic activation in the brain's reward pathway is one of the hallmark observations in depressed subjects. Would it come as a surprise to know that dopamine is also discovered to be functionally present and utilised in the retina portion of our eyes? Perhaps, not, if you can draw the connections between our eyes and brains in terms of reward sensitivity and positivity. There could even be potential sex differences in terms of the receptor type distribution in this sensory organ when you consider how men and women view aesthetic objects and the world of art differently. Does it come as a surprise that men tend to appreciate motor vehicles more than fine floral art? There are indeed dopaminergic intraretinal pathways in the back of our eyes. When depressed subjects viewed positive emotional images, there was impaired processing in the brain's nucleus accumbens region. Thus, you can see the connection between nurture by external stimuli and nature of the internal brain responses. It would be a sad miss to study them in isolation. In patients with Parkinson's disease, there is found a deficit in discriminating between colours and sensitivity towards contrast in their vision as they scored below healthy controls in the Fansworth-Munsell 100-Hue test, even in spite of being on medication (Büttner et al., 1995; Pieri et al., 2000; Price et al., 1992). Intuitively, it should be fine to say that the depletion of dopamine in Parkinson's disease is not exclusive to the substantia nigra region in the dorsal striatum of the brain, otherwise the retinal function of these patients would still be comfortably intact. This also points to the shared function between our eyes and brain that serves to promote pleasurable and positive reward processing beyond their specific isolated functions. Just imagine, for instance, what difference would it make to your level of delight or satisfaction between looking at a single deep red color rose and the various well-matched colours of a rainbow in the sky. Hint – colour discrimination plays a role. Or, is there a difference in your mood level between staring at a plain white lily and clusters of Bonica roses with varying shades of pink and white? Again, the function of contrast sensitivity comes into the picture. With dopamine, we get to enjoy the mood-elevating responses brought about by these two visual functions, which in turn ultimately serve to promote our enjoyment of the rich pleasurable properties of the world and universe around us. This is just one of the many examples of a demonstration of the holistic connection between different neurophysiological components of our body – in this case, our visual and brain systems, working in concert to ensure we get to enjoy the variety of beauty in life. At the end of the process of it all, we get to benefit, which is why there is a colorful and contrasted image of a curious squirrel on the cover of this book about dopamine! On the other hand, perhaps a promising future target for early diagnosis and treatment of Parkinson's disease could be to utilise these colour discrimination and contrast sensitivity tests for a more convenient, less invasive and inexpensively quicker detection of the disorder.

Figure 2: As a handy exercise, rate your level of positive experience in terms of arousal and pleasure when you view each of the above two images and you could easily estimate the level of dopamine release in response to the contrasting colour distinctiveness and diversity content of each image. How does your response to the image of the brighter cherry blossom petals on the left differ from the one on the right? It may be more interesting to compare such experimental outcomes between male and female individuals. Refer to Activity 1.0 for further experiment ideas.

Returning to the theme of clinical depression, we still have yet to see the correlation of the impaired dopamine function with long-term exposure to negative stimuli and past life experiences that could have produced dampened response to positive elements in an individual due to lasting modification of relevant neural circuits. Prospective studies are useful in this case for the study of children and adolescents at probable high risk for depression, based on variables such as disadvantaged living conditions and lack of social support, and to include these sociobehavioural measures with concurrent neuroscientific findings to better explore the temporal mediating factor in depression onset. On the other hand, the measure of level of positivity and negativity of stimuli and/or themes sample in studies is related to characteristic of valence, which entails emotional perception and evaluation, hence the brain's limbic region and associated pathways play an important part in the processing of emotional response by participants. Localized region and circuit connection abnormalities are equally significant in the disorder pathology (Downar et al., 2014; Zhang et al., 2011). A dysfunction in the limbic-cortical network along with reduced volume of certain prefrontal cortical regions have been discovered in depressed individuals (Bremner, 2002). It is also likely that responses to drug treatment can be modulated by abnormalities in limbic-subcortical network, thus emphasizing structural and functional aspects as additional challenging targets for treatment and the inadequate and temporal efficacy of standard medication. A question that could be raised is whether drug targets of neurotransmitter release produces a more transient or short-lived process that is subjected to daily fluctuations of both internal and external cues. It would be useful to draw a hierarchical stage-wise model of the process of depression beginning from the fundamental localized micro to region-wide macro levels. Nevertheless, it is difficult to determine the direction of effect which considers the role of neurotransmitter dysregulation, reduced dopaminergic activation and structural modification to answer the question of precedence in causality. Hence, it would be interesting to see studies with a more comprehensive design in methodology that attempt to combine multi-level concepts to benefit current understanding of dynamic neuronal interactions as opposed to examining them as discrete units or concepts.

Although dampened dopaminergic activity in response to positive stimuli likely contributes to the persistence of depressive mood in susceptible individuals, the exact causal mechanism which sets such anhedonic symptom in motion remains to be investigated in-depth and may not be a one-size-fits-all answer but entail a combination and integration of neural processes which span multiple levels of neuropsychological hierarchy. One may ask, given the knowledge of reduced dopaminergic transmission in the mesolimbic pathway, can dopamine agonists (ligands which bind to and activate dopaminergic receptors) be part of the future avenue of pharmacological treatment for clinical depression? The answer is that it is still too early to tell as we are not fully acquainted with the comprehensive list of possible side effects associated with a mere handful of agonist drugs being developed. As an example, although preclinical animal testing of the dopamine D1 receptor potentiator, DETQ, supports its potential use in the treatment of depression and related neuropsychiatric disorders, it also casts immunological and sleep-wake cycle concerns by raising histamine levels (Bruns et al., 2018). Thus, a profitable connection between feasibility in medical use and benefit-over-cost clinical outcome(s) would determine the progress in the future implementation of dopaminergic agonist drugs as intervention candidates for depressive disorder. Meantime, we could use some help from natural "psychoactive" comfort beverages, such as caffeine-containing coffee and hot chocolate as remedies. Caffeine, in coffee, is tied to antagonistic action on adenosine receptors, which in turn promotes the release of dopamine in the brain. As one would have guessed, excessive consumption of caffeine would eventually lead to heightened production of dopamine that culminates in temporary psychotic episodes. This is obviously a negative downside of utilising dopaminergic activity to relieve depression. On the other hand, a hot chocolate drink could likely bypass this channel by directly inducing the release of endorphins, a positive alternative to the involvement of dopamine, that could promote stress or depression relief by affecting our baseline emotions as well as reducing the depressive symptom of dysphoria (Castell, Pérez-Cano, & Bisson, 2013). In addition, there are considerable neuroprotective benefits associated with the numerous components present in cocoa and chocolate that help maintain our brain health and improve cognition (Madhavadas, Kapgal, Kutty, & Subramanian, 2016; Magrone, Russo,

& Jirillo, 2017; Marshall, 2007; Melzig, Putscher, Henklein, & Haber, 2000; Nehlig, 2013). Interestingly, considering the alleviating effect of dopamine on depression and its relation with cortisol, one could easily predict the outcome of antidepressants on cortisol levels. Antidepressants could increase plasma and salivary cortisol levels since the strategy of this class of medications is to counter "depressed" mood and lack of motivation for physical activity (Bhagwagar, Hafizi, & Cowen, 2002). This is an opposite function to antipsychotics, which in turn is the reverse of caffeine and dopamine agonists. The underlying denominator is the way these various neurochemical components affect the opposing patterns of expenditure and conservation of energy resource pool in the body, which could be traced to their impact on the immune system, a very important and significant system in human physiology. The effect of dopamine and cortisol is very comparable because they increase blood pressure most of the time, therefore they expend energy resource which in turn, is diverted away from the immune system. This helps explain why cortisone and similar substances dampen inflammation and autoimmunity since they work in opposition against the immune system by directing energy expending opportunity away from it. On the other hand, just as adenosine is a dopamine antagonising component, it works to conserve energy for use and support of the immune system. Adenosine promotes and induces sleep and the purpose of this strategy to allow major energy expending systems of the body to shut down during sleep in order to make way and channel energy resource for the immune system to work, increase and maintain production of its body defence cells and living components as well as waste removal at night. During the day, there is just too much utilization of energy that the immune system switches into a passive mode. This reminds of why antibiotics and related medications tend to be sleep inducing and we need additional amounts of rest during an illness. There is just no underestimation for the huge amounts of resources the immune system needs for cell production and functions during an infection. This is also one of the reasons why sleep is a necessity for humans and animals (Refer to Figure 2 for a concise breakdown of this discussion).

Effect on Attentional and Emotional Arousal

Increasing/Promoting   Decreasing/Dampening

Effect on Blood Pressure

Increasing   Decreasing

Effect on Immunoactivity

Decreasing   Increasing

Energy "Expender"   Energy "Conservant"

E.g. Caffeine, Dopamine   E.g. Adenosine, Antipsychotics

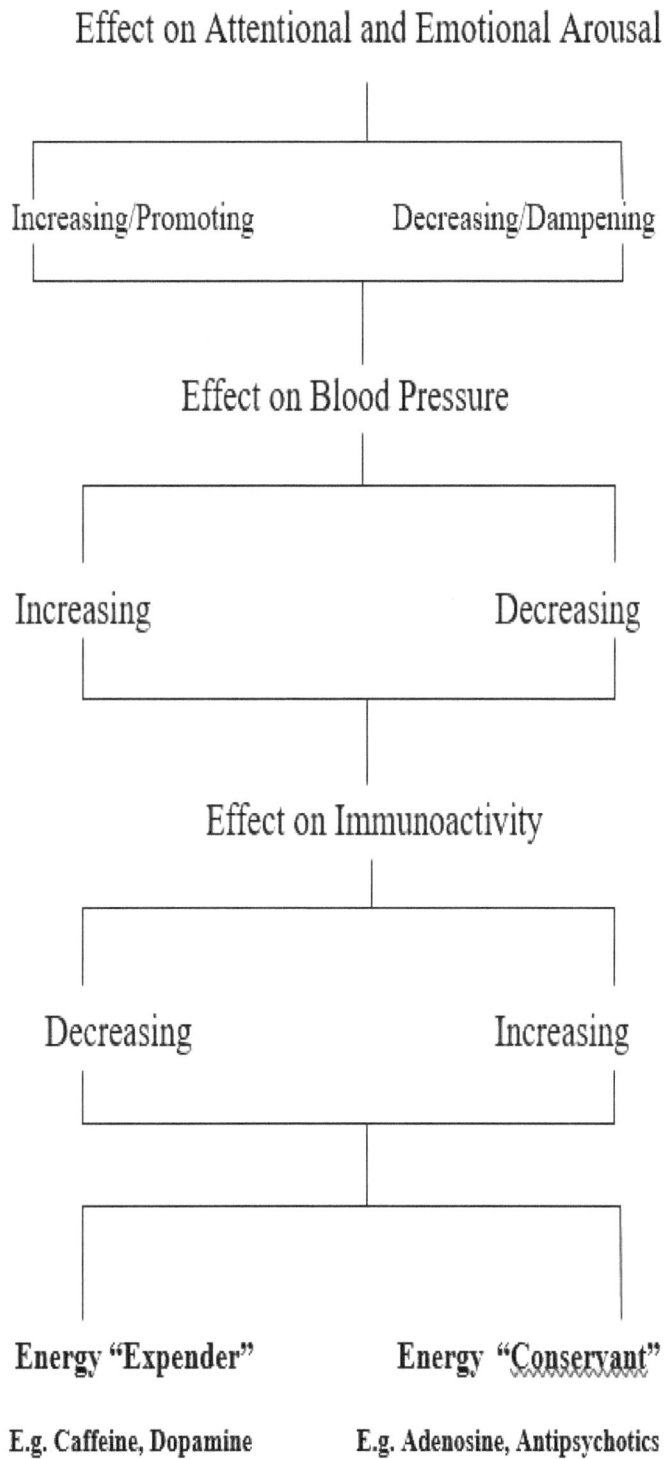

Figure 3: Breakdown of the opposing effects of neurochemicals on energy resource pool.

Chapter 6: The Effects of Stress on Dopamine Activity

With the significant roles which dopamine play on the disorders of clinical depression and schizophrenia, one may wonder how does one of life's greatest toils, that is stress, intertwine with these two debilitating yet rising causes for disability and further adds to their burdensome prognoses. No doubt, the effects, as in plurality, of stress are manifold. Upon looking at the numerous research documenting and analysing its effects on the levels of dopamine release, neurochemical activity and such, there is a reasonable conclusion attesting to the time-dependent alternating, or even self-opposing, pattern at play. This is where researchers looking at introducing confounding variables of addictive and stimulant drug components, e.g. cocaine, amphetamines, etc., into their animal or human subject studies missed it. If you want to genuinely look at the natural cycle of dopamine release, metabolism and replenishment within the human body and brain and how this paves the way to potential maladaptive pattern(s) like depressive and addiction disorders, it is best to avoid inclusion or administration of psychostimulant substances just so to obtain a higher power and statistically significant results for your studies. This attempt would not generate a better interpretation for your results even though you may be adding much data to support their statistical power. That said, there is a limited source which comprehensively and "clearly" details the time window sensitive, or biphasic and tonic, effects of stress on dopaminergic activity in the brain without the additional influence of psychostimulant drugs. However, that does not mean none. There are studies which successfully reported alternating initial increases followed by decreases in level of dopamine in the nucleus accumbens and ventral tegmental area (parts of the core mesolimbic reward circuit), as well as the medial prefrontal cortex, upon exposure to acute stress (Cabib, & Puglisi-Allegra, 2012, Kaneyuki et al., 1991).

The alternating increase-decrease pattern in dopamine levels and activity could account for the reason why some studies are reporting one way or the other, and that this is a time-sensitive effect. If we can look at the single overarching principle behind the purpose and existence of dopamine in our neurobiological makeup, it could give us a connecting-dots experience in understanding the goal of this self-opposing time-dependent pattern. Dopamine is the neurotransmitter which not only motivates, but also activates us, for the quest of obtaining both pleasure and positive relief from the negative circumstances in life. When acute stress seeks to introduce negative effect on our system, dopamine has to immediately spring into action to counter it with positivity, which explains why stress can induce psychotic symptoms, a "precursor" to schizophrenia (Mizrahi et al., 2012; Soliman et al., 2008). However, chronic stress linked to clinical depression likely proceeds from the "decreasing phase" of dopamine release (which could fall to extreme lows below basal level), which accounts for why depression tends to follow or co-occur with psychosis or schizophrenia. On the other hand, it could be argued that the subconscious goal of obtaining and self-administering addictive drugs, which leads to substance abuse disorder, is to compensate for an early-life or chronic low levels of endogenous dopamine release and related neurochemical activity. This compensatory mechanism, is of course, to ultimately increase dopamine release to way beyond its basal levels. In simple words, when at any instant, your brain's basal dopamine level is low or overused, you would naturally respond to this depletion by, for example, drinking coffee, which is an adenosine antagonist that will indirectly increase dopamine activity because adenosine tends to inhibit dopamine in certain brain regions. Therefore, the natural cycle is to increase whenever there is a decrease, and vice versa, just so to achieve a healthy balanced state that is neither in excess nor depletion. This could be a reason why coffee or caffeine has been reported to alleviate and decrease the risk of depression (Kanjanakorn, & Lee, 2017; Nehlig, 2016).

Chapter 7: Dopamine and the Cure for Substance Abuse Disorders

I once questioned a lecturer about whether people who have food addiction problems would also be less likely to engage in substance (illicit drugs) abuse since the pleasure of food consumption could raise dopamine levels in the brain as addictive drugs do, therefore substituting their inclination to approach these drugs just to experience the "highs"? The answer he gave me was quite negative and not very encouraging. I had hoped that food addiction could at least be a mild and harmless replacement for drugs of abuse. Although pleasure and comfort foods do not raise dopamine to overly extreme peak levels and neither supply the "addictive highs" that were difficult to resist, I wonder if a person who had ventured into the territory of food addiction prior to experimenting with such drugs, how would the outcome be different since he has an alternative go to remedy with built-in past experiences already etched in his memory? Needless to say, memories of addictive drug intake experience and an abuser's psychophysiological response(s) to it play a crucial and reinforcing role in substance abuse disorders. Just imagine one day, when a heavy drug user wakes up with amnesia and has apparently no recall of such drug-related memories in his mind, how would this affect his immediate and future potential cravings and reaction to the sight of a drug pill of which he already has completely no idea about. Sounds like a hit movie scenario, but it is still worth imagining the "positive" possibilities of it for a cure.

Returning to the discussion of dopamine, it cannot be undermined that most addictive drugs always end up affecting the levels of the powerful neurotransmitter dopamine at one of their end stages upon entry into the brain. Opioids, cannabis, amphetamines, cocaine – they all raise dopamine to considerable levels in the nucleus accumbens/ventral striatum. Well, perhaps cannabis is less so than the rest in the list and that is a reason why people have claimed that cannabis is not as dangerously addictive as most illicit drugs. This simply supports the fact that the greater the impact the drug has on dopamine, the greater its addictive potential. Such drug craving is multi-faceted which is subjected to the influence of emotional, cognitive, sensory (both internally and externally directed) and physical cue channels, in a similar way that the neurotransmitter dopamine helps process and oversee a diversity of these neural functions. No doubt, the very stage of psychostimulant drug craving induces an increase in dopamine release, which is interestingly associated with systolic blood pressure increase and the HPA-axis, emphasising the under-recognised global physiological role of dopamine in these systems (Grace, 2000; Sinha et al., 2003; Volkow et al., 2008). Dopamine can have a blood pressure increasing effect and for individuals with substance abuse disorders who are undergoing treatment, the motivation to maintain a longer duration of abstinence from the drug can be another source of stress inducer that triggers the activation of the HPA-axis.

Whenever there is a targeted inhibition on the natural or spontaneous flow of neurophysiological processes taking place within an individual, there is an opposing homeostatic strategy being triggered, just as Newton's third law of motion in physics describes how there is an equal and opposite reaction to every action. This could be a potential explanation to why our blood pressure tends to increase to oppose any psychophysiological stress directed against us in the event that they are happening. It is wondered whether antipsychotic and blood pressure lowering medications can be utilised to reduce the early phase negative effects of drug craving and promote blood vessel relaxation, thereby easing this unpleasant stage of recurring substance abuse problem. Drug craving is drug-specific and cue-specific that are moderated by individual past experiences. However, even in the absence of visual and tangible cue(s), memory and reminiscence of past drug-taking experiences can be a potent internal trigger that emerges any moment after a period of abstinence and extinction. Therefore, both internal and external sources of trigger of drug craving are equally significant despite the major research focus on physical and tangible cues.

There are drug rehabilitation programmes which aim to induce the negative aversive psychophysiological responses to a drug in an abuser after consumption in order to deter him from relapsing into drug abuse again. However, this strategy is less likely to succeed because ultimately, the positive pleasurable memories related to drug intake experiences would take over the negative ones, nullifying their short-lived negative affects because, quite simply, the negative aversive responses do not elevate dopamine levels, hence, do not induce formation of sufficient long-term memories in the brain. It is a wonder, whether a novel strategy for curing addiction could be devising a method of bypassing the drug intake pathway or entry into the brain in such a way that the drug abuser would not need the harmful substance to raise dopamine levels in the nucleus accumbens, because ultimately, dopamine could be responsible for inducing the major pleasures and positive experiences related to drug-taking. Could the ideal candidate be L-DOPA, a dopamine precursor that can bypass the blood-brain-barrier, and has been used for treatment of Parkinson's disease? On the other hand, would this strategy only serve to increase the intensity of craving and distressing withdrawal symptoms after more dopamine is being introduced into the system? This is a challenging and complex discussion.

Most researchers are already aware by now of the elevated health risks associated with addictive drug taking during the vulnerable age window of adolescence. Why is this so? Intuitively, a part of the answer may lie in the crucial neurophysiological and hormonal changes and transformation that are taking place at such stage of growth. In a review by Kennedy (2018), it is reported that the activity of drug metabolizing enzymes appears to be reduced as one nears prepuberty, particularly for girls, whereas such enzymes work at highly elevated levels during childhood. It is interesting to note that during this phase of decreased activity of drug metabolism when a lesser amount of psychoactive drugs is being efficiently metabolized, detoxified and excreted from the body after intake, the level of female reproductive hormone, estrogen, increases. Research with caffeine and related substances support the observation of reduced function of drug metabolizing enzymes during adolescence. There is likely a regulation of allocation and distribution of energy resources involved between reproductive hormonal changes and drug metabolism enzymatic needs in which the former is prioritised and privileged over the latter. It is therefore an important stage of tradeoffs when the winner is already predetermined but which would also be in effect only for a certain period. In this case, the period of pubertal adolescence is surely the worst time to choose to experiment with addictive drugs or substance abuse if such substances are less likely to be efficiently metabolized by the hepatic enzymes that they remain in the body for an extended duration of time. Drug overdose will therefore be an elevated risk for teens and adolescents more than older adults with the resulting complications that may arise from resource competition with rapidly maturing reproductive hormonal system potentially taxing the body.

Activity 1.0: Visual Response to Colour Properties

Let us revert back to Figure 2 of Chapter 5 and analyse the underlying individual colour properties of these two contrast flower images using an online TinEye Labs colour extractor powered by MulticolorEngine (© 2021 TinEye) which categorises each coded shade of colours found in an image according to their proportion percentages detected.

Image 1

Extracted color palette

Color map regions   Proportional palette

27.3 %	#cf9bb6	Pink
27.0 %	#31481d	Green
25.3 %	#9c9189	Grey
11.0 %	#272b1f	Black
2.8 %	#b35c67	Pink
2.4 %	#e3709e	Pink
1.6 %	#664846	Brown
1.1 %	#cc454c	Red
0.9 %	#7c5835	Brown
0.5 %	#682d20	Brown

Source image

Figure 4: Extracted colour palette results by TinEye Labs for Image 1.

Points for Thought

If your ratings for arousal and pleasure is high for the above cherry blossom flower petals image, there may be a correlational relationship with its colour distinctiveness and diversity properties as analysed by the colour extractor and presented in the screenshot Figure 4. The colour distinctiveness/variability measure can be represented by the greater number of individual colour shades, i.e. there are 10 recognisable distinct colours and subcolours (e.g. 3 different pink colour shades/subcolours). If you will compare them with the second extracted colour palette for Image 2 in Figure 5 below, what can you conclude about their relationship(s) with possible dopamine responses in your brain to these 2 visual images? And, is there a likelihood that the arousal and pleasure ratings between male and female individuals would show a notable discrepancy and opposing views, with men's ratings between the two images exhibiting a smaller window of difference than women's ratings? As you can see, colour diversity is greater in Image 1 and response by the dopamine receptors to a greater variation in colour wavelength stimuli which induce them may be responsible for our overall above baseline levels of emotional arousal and pleasure. This implies a highly integrated global neuronal connectivity between our primary visual and visual association cortices and mesolimbic reward pathways which "inter-processes" at great speed, even within seconds.

Image 2

Extracted color palette

Color map regions

Proportional palette

Source image

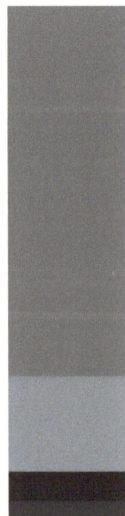

72.2 %	#7e7d75	Grey
18.6 %	#8d9198	Grey
5.6 %	#2b2f25	Grey
3.0 %	#3c413e	Grey
0.5 %	#475e2f	Green

Exclude background color from extracted colors

Exclude interior background color from extracted colors

Figure 5: Extracted colour palette results by TinEye Labs for Image 2.

References

Adams Jr., J. D. (2012). Parkinson's disease - apoptosis and dopamine oxidation. *Open Journal of Apoptosis, 1(1),* 1-8. doi:10.4236/ojapo.2012.11001

Akkaya, C., Kaya, B., Kotan, Z., Sarandol, A., Ersoy, C., & Kirli, S. (2009). Hyperpro-lactinemia and possibly related development of prolactinoma during amisulpride treatment; three cases. Journal of Psychopharmacology, 23(6), 723-726.

Armentero, M. T., Pinna, A., Ferré, S., Lanciego, J. L., Müller, C. E., & Franco, R. (2011). Past, present and future of A(2A) adenosine receptor antagonists in the therapy of Parkinson's disease. *Pharmacology and Therapeutics, 132,* 280–299. doi:10.1016/j.pharmthera.2011.07.004

Backhaus, J., Junghanns, K., Born, J., Hohaus, K., Faasch, F., & Hohagen, F. (2006). Impaired declarative memory consolidation during sleep in patients with primary insomnia: influence of sleep architecture and nocturnal cortisol release. *Biological psychiatry*, *60*(12), 1324-1330.

Beck, G. C., Brinkkoetter, P., Hanusch, C., Schulte, J., van Ackern, K., van der Woude, F. J., & Yard, B. A. (2004). Clinical review: immunomodulatory effects of dopamine in general inflammation. *Critical Care*, *8*(6), 485.

Benedetti, M. D., Bower, J. H., Maraganore, D. M., McDonnell, S. K., Peterson, B. J., Ahlskog, J. E., ... & Rocca, W. A. (2000). Smoking, alcohol, and coffee consumption preceding Parkinson's disease: a case-control study. *Neurology*, *55*(9), 1350-1358.

Beninger, R. J. (1983). The role of dopamine in locomotor activity and learning. *Brain Research Reviews*, *6*(2), 173-196.

Bhagwagar, Z., Hafizi, S., & Cowen, P. (2002). Acute citalopram administration produces correlated increases in plasma and salivary cortisol. *Psychopharmacology*, *163*(1), 118-120.

Borota, D., Murray, E., Keceli, G., Chang, A., Watabe, J. M., Ly, M., ... & Yassa, M. A. (2014). Post-study caffeine administration enhances memory consolidation in humans. Nature neuroscience, 17(2), 201-203.

Brandies, R., & Yehuda, S. (2008). The possible role of retinal dopaminergic system in visual performance. *Neuroscience & Biobehavioral Reviews, 32*(4), 611-656.

Bremner, J. D., Vythilingam, M., Vermetten, E., Nazeer, A., Adil, J., Khan, S., ... & Charney, D. S. (2002). Reduced volume of orbitofrontal cortex in major depression. *Biological psychiatry, 51*(4), 273-279.

Bruns, R. F., Mitchell, S. N., Wafford, K. A., Harper, A. J., Shanks, E. A., Carter, G., ... & Beck, J. P. (2018). Preclinical profile of a dopamine D1 potentiator suggests therapeutic utility in neurological and psychiatric disorders. *Neuropharmacology, 128*, 351-365.

Büttner, T., Kuhn, W., Müller, T., Patzold, T., Heidbrink, K., & Przuntek, H. (1995). Distorted color discrimination in 'de nova' parkinsonian patients. *Neurology, 45*(2), 386-387.

Cabib, S., & Puglisi-Allegra, S. (2012). The mesoaccumbens dopamine in coping with stress. *Neuroscience & Biobehavioral Reviews, 36*(1), 79-89.

Carboni, E., Imperato, A., Perezzani, L., & Di Chiara, G. (1989). Amphetamine, cocaine, phencyclidine and nomifensine increase extracellular dopamine concentrations preferentially in the nucleus accumbens of freely moving rats. *Neuroscience, 28*(3), 653-661.

Carlson, N. R. (2011). *Foundations of behavioral neuroscience.* Boston, MA: Allyn and Bacon.

Castell, M., Pérez-Cano, F. J., & Bisson, J. F. (2013). Clinical benefits of cocoa: An overview. In *Chocolate in health and nutrition* (pp. 265-275). Humana Press, Totowa, NJ.

Checkoway, H., Powers, K., Smith-Weller, T., Franklin, G. M., Longstreth Jr, W. T., & Swanson, P. D. (2002). Parkinson's disease risks associated with cigarette smoking, alcohol consumption, and caffeine intake. *American journal of epidemiology, 155*(8), 732-738.

Chen, G. H., Xia, L., Wang, F., LI, X. W., & Jiao, C. A. (2016). Patients with chronic insomnia have selective impairments in memory that are modulated by cortisol. *Psychophysiology, 53*(10), 1567-1576.

Chen, J., Xu, K., Petzer, J. P., Staal, R., Xu, Y., Beilstein, M.,...Schwarzschild, M. A. (2001). Neuroprotection by caffeine and A2A adenosine receptor inactivation in a model of Parkinson's disease. *The Journal of Neuroscience, 21,* 1-6.

Delgado, P. L., Charney, D. S., Price, L. H., Landis, H., & Heninger, G. R. (1989). Neuroendocrine and behavioral effects of dietary tryptophan restriction in healthy subjects. *Life sciences, 45*(24), 2323-2332.

Downar, J., Geraci, J., Salomons, T. V., Dunlop, K., Wheeler, S., McAndrews, M. P., ... & Flint, A. J. (2014). Anhedonia and reward-circuit connectivity distinguish nonresponders from responders to dorsomedial prefrontal repetitive transcranial magnetic stimulation in major depression. *Biological psychiatry, 76*(3), 176-185.

Dunlop, B. W., & Nemeroff, C. B. (2007). The role of dopamine in the pathophysiology of depression. *Archives of general psychiatry, 64*(3), 327-337.

Duszkiewicz, A. J., McNamara, C. G., Takeuchi, T., & Genzel, L. (2019). Novelty and dopaminergic modulation of memory persistence: a tale of two systems. *Trends in neurosciences, 42*(2), 102-114.

Dutcher, J. M., Creswell, J. D., Pacilio, L. E., Harris, P. R., Klein, W. M. P., Levine, J. M.,...Eisenberger, N. I. (2016). Self-affirmation activates the ventral striatum: A possible reward-related mechanism for self-affirmation. *Psychological Science.* Advance online publication. doi: 10.1177/0956797615625989

Epstein, J., Pan, H., Kocsis, J. H., Yang, Y., Butler, T., Chusid, J., ... & Silbersweig, D. A. (2006). Lack of ventral striatal response to positive stimuli in depressed versus normal subjects. *American Journal of Psychiatry, 163*(10), 1784-1790.

Flagel, S. B., Clark, J. J., Robinson, T. E., Mayo, L., Czuj, A., Willuhn, I., ... & Akil, H. (2011). A selective role for dopamine in stimulus–reward learning. *Nature, 469*(7328), 53.

Foley, C., Corvin, A., & Nakagome, S. (2017). Genetics of schizophrenia: ready to translate?. *Current Psychiatry Reports, 19*(9), 61.

Gasbarri, A., Verney, C., Innocenzi, R., Campana, E., & Pacitti, C. (1994). Mesolimbic dopaminergic neurons innervating the hippocampal formation in the rat: a combined retrograde tracing and immunohistochemical study. *Brain research, 668*(1-2), 71-79.

Grace, A. A. (2000). The tonic/phasic model of dopamine system regulation and its implications for understanding alcohol and psychostimulant craving. *Addiction, 95*(8s2), 119-128.

Greenamyre, J. T. (2001). Glutamatergic influences on the basal ganglia. *Clinical Neuropharmacology, 24,* 65–70.

Halliday, G. M., & Stevens, C. H. (2011). Glia: Initiators and progressors of pathology in Parkinson's disease. *Movement Disorders, 26,* 6–17. doi: 10.1002/mds.23455

Hamidovic, A., Childs, E., Conrad, M., King, A., & de Wit, H. (2010). Stress-induced changes in mood and cortisol release predict mood effects of amphetamine. *Drug and alcohol dependence, 109*(1-3), 175-180.

Henriksen, M. G., Nordgaard, J., & Jansson, L. B. (2017). Genetics of schizophrenia: overview of methods, findings and limitations. *Frontiers in human neuroscience, 11,* 322.

Hirsch, E. C., & Hunot, S. (2009). Neuroinflammation in Parkinson's disease: A target for neuroprotection? *Lancet Neurology, 8,* 382–397.

Hunter, M. D., Mysorekar, S., & Woodruff, P. W. R. (2005). Visual broadcast in schizophrenia. *Medical humanities, 31*(1), 55-55.

Jose, P. A., Eisner, G. M., & Felder, R. A. (2002). Role of dopamine receptors in the kidney in the regulation of blood pressure. *Current opinion in nephrology and hypertension, 11*(1), 87-92.

Kaneyuki, H., Yokoo, H., Tsuda, A., Yoshida, M., Mizuki, Y., Yamada, M., & Tanaka, M. (1991). Psychological stress increases dopamine turnover selectively in mesoprefrontal dopamine neurons of rats: reversal by diazepam. *Brain research, 557*(1-2), 154-161.

Kanjanakorn, A., & Lee, J. (2017). Examining emotions and comparing the EsSense Profile® and the Coffee Drinking Experience in coffee drinkers in the natural environment. *Food quality and preference, 56,* 69-79.

Kennedy, M. J. (2008). Hormonal regulation of hepatic drug-metabolizing enzyme activity during adolescence. *Clinical Pharmacology & Therapeutics, 84*(6), 662-673.

Kong, J., Shepel, N., Holden, C. P., Mackiewicz, M., Pack, A. I., & Geiger, J. D. (2002). Brain glycogen decreases with increased periods of wakefulness: Implications for homeostatic drive to sleep. *Journal of Neuroscience, 22,* 5581-5587.

Kudoh, M., & Shibuki, K. (2006). Sound sequence discrimination learning motivated by reward requires dopaminergic D2 receptor activation in the rat auditory cortex. *Learning & Memory, 13*(6), 690-698.

Lancelot, E., & Beal, M. F. (1998). Glutamate toxicity in chronic neurodegenerative disease. *Progress in Brain Research, 116,* 331–347.

Lemon, N., & Manahan-Vaughan, D. (2006). Dopamine D1/D5 receptors gate the acquisition of novel information through hippocampal long-term potentiation and long-term depression. *Journal of Neuroscience, 26*(29), 7723-7729.

Leyton, M., Boileau, I., Benkelfat, C., Diksic, M., Baker, G., & Dagher, A. (2002). Amphetamine-induced increases in extracellular dopamine, drug wanting, and novelty seeking: a PET/[11C] raclopride study in healthy men. *Neuropsychopharmacology, 27*(6), 1027-1035.

Li, S., Cullen, W. K., Anwyl, R., & Rowan, M. J. (2003). Dopamine-dependent facilitation of LTP induction in hippocampal CA1 by exposure to spatial novelty. *Nature neuroscience, 6*(5), 526-531.

Litteljohn, D., Mangano, E., Clarke, M., Bobyn, J., Moloney, K., & Hayley, S. (2010). Inflammatory mechanisms of neurodegeneration in toxin-based models of Parkinson's disease. *Parkinson's Disease, 2011,* 1-18. doi:10.4061/2011/713517

Locke, C. J., Fox, S. A., Caldwell, G. A., & Caldwell, K. A. (2008). Acetaminophen attenuates dopamine neuron degeneration in animal models of Parkinson's disease. *Neuroscience Letters, 2,* 129-133. doi:10.1016/j.neulet.2008.05.003

Lopes, L. V., Sebastião, A. M., & Ribeiro, J. A. (2011). Adenosine and related drugs in brain diseases: Present and future in clinical trials. *Current Topics in Medicinal Chemistry, 11,* 1087–1101.

Maddock, R. J., Garrett, A. S., & Buonocore, M. H. (2003). Posterior cingulate cortex activation by emotional words: fMRI evidence from a valence decision task. *Human brain mapping, 18*(1), 30-41.

Madhavadas, S., Kapgal, V. K., Kutty, B. M., & Subramanian, S. (2016). The neuroprotective effect of dark chocolate in monosodium glutamate-induced nontransgenic Alzheimer disease model rats: biochemical, behavioral, and histological studies. *Journal of dietary supplements, 13*(4), 449-460.

Magrone, T., Russo, M. A., & Jirillo, E. (2017). Cocoa and dark chocolate polyphenols: from biology to clinical applications. *Frontiers in immunology*, *8*, 677.

Marshall, S. (2007). Chocolate: indulgence or medicine?. *Pharmaceutical Journal (Vol 278)*.

McFarland, B. R., & Klein, D. N. (2009). Emotional reactivity in depression: diminished responsiveness to anticipated reward but not to anticipated punishment or to non-reward or avoidance. *Depression and anxiety*, *26*(2), 117-122.

McKenna, F., McLaughlin, P. J., Lewis, B. J., Sibbring, G. C., Cummerson, J. A., Bowen-Jones, D., & Moots, R. J. (2002). Dopamine receptor expression on human T-and B-lymphocytes, monocytes, neutrophils, eosinophils and NK cells: a flow cytometric study. *Journal of neuroimmunology*, *132*(1-2), 34-40.

McNamara, C. G., Tejero-Cantero, Á., Trouche, S., Campo-Urriza, N., & Dupret, D. (2014). Dopaminergic neurons promote hippocampal reactivation and spatial memory persistence. *Nature neuroscience*, *17*(12), 1658-1660.

Melzig, M. F., Putscher, I., Henklein, P., & Haber, H. (2000). In vitro pharmacological activity of the tetrahydroisoquinoline salsolinol present in products from Theobroma cacao L. like cocoa and chocolate. *Journal of ethnopharmacology*, *73*(1-2), 153-159.

Menezes, J., Alves, N., Borges, S., Roehrs, R., de Carvalho Myskiw, J., Furini, C. R. G., ... & Mello-Carpes, P. B. (2015). Facilitation of fear extinction by novelty depends on dopamine acting on D1-subtype dopamine receptors in hippocampus. *Proceedings of the National Academy of Sciences*, *112*(13), E1652-E1658.

Mennini, T., Mocaer, E., & Garattini, S. (1987). Tianeptine, a selective enhancer of serotonin uptake in rat brain. *Naunyn-Schmiedeberg's archives of pharmacology*, *336*(5), 478-482.

Meyer, J. S., & Quenzer, L. F. (2013). Psychopharmacology: Drugs, the brain, and behavior. Sinauer Associates.

Mizrahi, R., Addington, J., Rusjan, P. M., Suridjan, I., Ng, A., Boileau, I., ... & Wilson, A. A. (2012). Increased stress-induced dopamine release in psychosis. *Biological psychiatry*, *71*(6), 561-567.

Mondelli, V., Dazzan, P., Hepgul, N., Di Forti, M., Aas, M., D'Albenzio, A., ... & Morgan, C. (2010). Abnormal cortisol levels during the day and cortisol awakening response in first-episode psychosis: the role of stress and of antipsychotic treatment. *Schizophrenia research*, *116*(2-3), 234-242.

Morale, M. C., Serra, P. A., L'Episcopo, F., Tirolo, C., Caniglia, S., Testa, N., ... & Miele, E. (2006). Estrogen, neuroinflammation and neuroprotection in Parkinson's disease: glia dictates resistance versus vulnerability to neurodegeneration. *Neuroscience*, *138*(3), 869-878.

Morelli, M., Carta, A. R., Kachroo, A., & Schwarzschild, M. A. (2010). Pathophysiological roles for purines: Adenosine, caffeine and urate. *Progress in Brain Research*, *183*, 183–208. doi:10.1016/S0079-6123(10)83010-9

Morissette, M., Al Sweidi, S., Callier, S., & Di Paolo, T. (2008). Estrogen and SERM neuroprotection in animal models of Parkinson's disease. *Molecular and cellular endocrinology*, *290*(1-2), 60-69.

Morrison, A. P., Frame, L., & Larkin, W. (2003). Relationships between trauma and psychosis: A review and integration. *British Journal of Clinical Psychology*, *42*(4), 331-353.

Müller, N., Weidinger, E., Leitner, B., & Schwarz, M. J. (2015). The role of inflammation in schizophrenia. *Frontiers in neuroscience*, *9*, 372.

Murphy, F. C., Michael, A., Robbins, T. W., & Sahakian, B. J. (2003). Neuropsychological impairment in patients with major depressive disorder: the effects of feedback on task performance. *Psychological medicine, 33*(03), 455-467.

Myburgh JA, Upton RN, Grant C, Martinez A. A comparison of the effects of norepinephrine, epinephrine, and dopamine on cerebral blood flow and oxygen utilisation.

Nau, R., Sander, D., & Klingelhöfer, J. (1992). Relationships between dopamine infusions and intracranial hemodynamics in patients with raised intracranial pressure. *Clinical neurology and neurosurgery, 94*(2), 143-148.

Nehlig, A. (2013). The neuroprotective effects of cocoa flavanol and its influence on cognitive performance. *British journal of clinical pharmacology, 75*(3), 716-727.

Nehlig, A. (2016). Effects of coffee/caffeine on brain health and disease: What should I tell my patients?. *Practical neurology, 16*(2), 89-95.

Nestler, E. J., & Carlezon, W. A. (2006). The mesolimbic dopamine reward circuit in depression. *Biological psychiatry, 59*(12), 1151-1159.

Perroud, N., & Huguelet, P. (2004). A possible effect of amisulpride on a prolactinoma growth in a woman with borderline personality disorder. Pharmacological research, 50(3), 377-379.

Pieri, V., Diederich, N. J., Raman, R., & Goetz, C. G. (2000). Decreased color discrimination and contrast sensitivity in Parkinson's disease. *Journal of the neurological sciences, 172*(1), 7-11.

Pizzagalli, D. A., Holmes, A. J., Dillon, D. G., Goetz, E. L., Birk, J. L., Bogdan, R., ... & Fava, M. (2009). Reduced caudate and nucleus accumbens response to rewards in unmedicated individuals with major depressive disorder. *American Journal of Psychiatry, 166*(6), 702-710.

Pontieri, F. E., Tanda, G., & Di Chiara, G. A. E. T. A. N. O. (1995). Intravenous cocaine, morphine, and amphetamine preferentially increase extracellular dopamine in the" shell" as compared with the" core" of the rat nucleus accumbens. *Proceedings of the National Academy of Sciences*, *92*(26), 12304-12308.

Popoli, P., Betto, P., Reggio, R., & Ricciarello, G. (1995). Adenosine A(2A) receptor stimulation enhances striatal extracellular glutamate levels in rats. *European Journal of Pharmacology, 287*, 215–217.

Price, M. J., Feldman, R. G., Adelberg, D., & Kayne, H. (1992). Abnormalities in color vision and contrast sensitivity in Parkinson's disease. *Neurology*, *42*(4), 887-887.

Ract, C., Vigué, B., Bodjarian, N., Mazoit, J. X., Samii, K., & Tadié, M. (2001). Comparison of dopamine and norepinephrine after traumatic brain injury and hypoxic-hypotensive insult. *Journal of neurotrauma*, *18*(11), 1247-1254.

Ravizza, T., Galanopoulou, A. S., Velıšková, J., & Moshé, S. L. (2002). Sex differences in androgen and estrogen receptor expression in rat substantia nigra during development: an immunohistochemical study. *Neuroscience*, *115*(3), 685-696.

Reale, M., Iarlori, C., Thomas, A., Gambi, D., Perfetti, B., Di Nicola, M., & Onofrj, M. (2009). Peripheral cytokines profile in Parkinson's disease. *Brain, Behavior, and Immunity, 23*, 55–63. doi:10.1016/j.bbi.2008.07.003

Rodenbeck, A., Huether, G., Rüther, E., & Hajak, G. (2002). Interactions between evening and nocturnal cortisol secretion and sleep parameters in patients with severe chronic primary insomnia. *Neuroscience letters*, *324*(2), 159-163.

Roth, T., Roehrs, T., & Pies, R. (2007). Insomnia: pathophysiology and implications for treatment. *Sleep medicine reviews*, *11*(1), 71-79.

Rothschild, A. J., Langlais, P. J., Schatzberg, A. F., Walsh, F. X., Cole, J. O., & Bird, E. D. (1984). Dexamethasone increases plasma free dopamine in man. *Journal of psychiatric research*, *18*(3), 217-223.

Ruhé, H. G., Mason, N. S., & Schene, A. H. (2007). Mood is indirectly related to serotonin, norepinephrine and dopamine levels in humans: a meta-analysis of monoamine depletion studies. *Molecular psychiatry, 12*(4), 331.

Schuff, K. G., Hentges, S. T., Kelly, M. A., Binart, N., Kelly, P. A., Iuvone, P. M., ... & Low, M. J. (2002). Lack of prolactin receptor signaling in mice results in lactotroph proliferation and prolactinomas by dopamine-dependent and-independent mechanisms. The Journal of clinical investigation, 110(7), 973-981.

Schultz, W. (2015). Neuronal reward and decision signals: from theories to data. *Physiological Reviews, 95*(3), 853-951.

Schwarzschild, M. A., Chen, J. F., Tennis, M., Messing, S., Kamp, C., Ascherio, A., Holloway, R. G., Marek, K., Tanner, C. M., McDermott, M., Lang, A. E., & The Parkinson Study Group. (2003). Relating caffeine consumption to Parkinson's disease progression and dyskinesias development. *Movement Disorders, 18,* 1082–1083. doi:10.1002/mds.10585

Shinkai, S., Watanabe, S., Asai, H., & Shek, P. N. (1996). Cortisol response to exercise and post-exercise suppression of blood lymphocyte subset counts. *International journal of sports medicine, 17*(08), 597-603.

Simola, N., Morelli, M., & Carta, A. R. (2007). The 6-Hydroxydopamine model of Parkinson's disease. *Neurotoxicity Research, 11,* 151-167.

Sinha, R., Talih, M., Malison, R., Cooney, N., Anderson, G. M., & Kreek, M. J. (2003). Hypothalamic-pituitary-adrenal axis and sympatho-adreno-medullary responses during stress-induced and drug cue-induced cocaine craving states. *Psychopharmacology, 170*(1), 62-72.

Soliman, A., O'driscoll, G. A., Pruessner, J., Holahan, A. L. V., Boileau, I., Gagnon, D., & Dagher, A. (2008). Stress-induced dopamine release in humans at risk of psychosis: a [11 C] raclopride PET study. *Neuropsychopharmacology, 33*(8), 2033-2041.

Sperling, S., & Bhatt, H. (2016). Prolactinoma: A Massive Effect on Bone Mineral Density in a Young Patient. Case reports in endocrinology, 2016.

Szarfman, A., Tonning, J. M., Levine, J. G., & Doraiswamy, P. M. (2006). Atypical antipsychotics and pituitary tumors: a pharmacovigilance study. Pharmacotherapy: The Journal of Human Pharmacology and Drug Therapy, 26(6), 748-758.

Tian, N., Xu, H.P., & Wang P. (2015). Dopamine D2 receptors preferentially regulate the development of light responses of the inner retina. *European Journal of Neuroscience,* 41(1),17-30.

Tran, A. H., Uwano, T., Kimura, T., Hori, E., Katsuki, M., Nishijo, H., & Ono, T. (2008). Dopamine D1 receptor modulates hippocampal representation plasticity to spatial novelty. *Journal of Neuroscience, 28*(50), 13390-13400.

Tremblay, L. K., Naranjo, C. A., Graham, S. J., Herrmann, N., Mayberg, H. S., Hevenor, S., & Busto, U. E. (2005). Functional neuroanatomical substrates of altered reward processing in major depressive disorder revealed by a dopaminergic probe. *Archives of general psychiatry, 62*(11), 1228-1236.

Tripathy, D., & Grammas, P. (2009). Acetaminophen protects brain endothelial cells against oxidative stress. *Microvascular Research, 77,* 289-296. doi:10.1016/j.mvr.2009.02.002

Tripathy, D., Sanchez, A., Yin, X., Martinez, J., & Grammas, P. (2012). Age-related decrease in cerebrovascular-derived neuroprotective proteins: Effect of acetaminophen. *Microvascular Research, 84,* 278-285. doi:10.1016/j.mvr.2012.08.004

Van Den Eeden, S. K., Tanner, C. M., Bernstein, A. L., Fross, R. D., Leimpeter, A., Bloch, D. A., & Nelson, L. M. (2003). Incidence of Parkinson's disease:

variation by age, gender, and race/ethnicity. *American journal of epidemiology, 157*(11), 1015-1022.

Vaughan, C. J., Aherne, A. M., Lane, E., Power, O., Carey, R. M., & O'Connell, D. P. (2000). Identification and regional distribution of the dopamine D1A receptor in the gastrointestinal tract. *American Journal of Physiology-Regulatory, Integrative and Comparative Physiology, 279*(2), R599-R609.

Venkatasubramanian, G., Chittiprol, S., Neelakantachar, N., Shetty, T., & Gangadhar, B. N. (2010). Effect of antipsychotic treatment on Insulin-like Growth Factor-1 and cortisol in schizophrenia: a longitudinal study. *Schizophrenia research, 119*(1-3), 131-137.

Volkow, N. D., Wang, G. J., Telang, F., Fowler, J. S., Logan, J., Childress, A. R., ... & Wong, C. (2008). Dopamine increases in striatum do not elicit craving in cocaine abusers unless they are coupled with cocaine cues. *Neuroimage, 39*(3), 1266-1273.

Wand, G. S., Oswald, L. M., McCaul, M. E., Wong, D. F., Johnson, E., Zhou, Y., ... & Kumar, A. (2007). Association of amphetamine-induced striatal dopamine release and cortisol responses to psychological stress. *Neuropsychopharmacology, 32*(11), 2310-2320.

Winterer, G., & Weinberger, D. R. (2004). Genes, dopamine and cortical signal-to-noise ratio in schizophrenia. *Trends in neurosciences, 27*(11), 683-690.

Wise, R. A. (2004). Dopamine, learning and motivation. *Nature reviews neuroscience, 5*(6), 483.

Wittmann, B. C., Bunzeck, N., Dolan, R. J., & Düzel, E. (2007). Anticipation of novelty recruits reward system and hippocampus while promoting recollection. *Neuroimage, 38*(1), 194-202.

Wyvell, C. L., & Berridge, K. C. (2000). Intra-accumbens amphetamine increases the conditioned incentive salience of sucrose reward: enhancement of reward "wanting" without enhanced "liking" or response reinforcement. *Journal of Neuroscience*, *20*(21), 8122-8130.

Xing, F., Hu, X., Jiang, J., Ma, Y., & Tang, A. (2016). A meta-analysis of low-dose dopamine in heart failure. International journal of cardiology, 222, 1003-1011.

Zhang, Y., Chen, Y., Wu, J., Manaenko, A., Yang, P., Tang, J., ... & Zhang, J. H. (2015). Activation of dopamine D2 receptor suppresses neuroinflammation through αB-crystalline by inhibition of NF-κB nuclear translocation in experimental ICH mice model. Stroke, 46(9), 2637-2646.

Zhang, J., Liu, X., Lei, X., Wang, L., Guo, L., Zhao, G., & Lin, G. (2010). Discovery and synthesis of novel luteolin derivatives as DAT agonists. *Bioorganic & medicinal chemistry*, *18*(22), 7842-7848.

Zhang, J., Wang, J., Wu, Q., Kuang, W., Huang, X., He, Y., & Gong, Q. (2011). Disrupted brain connectivity networks in drug-naive, first-episode major depressive disorder. *Biological psychiatry*, *70*(4), 334-342.

www.ingramcontent.com/pod-product-compliance
Lightning Source LLC
Chambersburg PA
CBHW061140030426
42335CB00002B/61